The Dragons Counting Series

Cici Counts by Ones

0 - 10 Forward

written by

Naomi Glasarth

illustrated by

Danh Tran

a burning butterfly book

Cici Counts by Ones, 1 - 10 Forward Copyright © 2017 Naomi Glasarth
All rights reserved.

Cover, layout, and illustrations copyright © 2017 by Burning Butterfly Books
Illustrations by Danh Tran Art and are created digitally.

Library of Congress Control Number: 2017954716

Hardback ISBN	10	1-947344-98-6
	13	978-1-947344-98-3
Paperback ISBN	10	1-947344-91-9
	13	978-1-947344-91-4
eBook ISBN	10	1-947344-97-8
	13	978-1-947344-97-6

Published by Burning Butterfly Books
Camp Verde, Arizona
www.burningbutterflybooks.com

For Cyndi.

Words to Know

flowers	garden
empty	spade
dig	holes
fingers	plant
seeds	wait
grow	tall
tasty	giant
dinner	plates
dance	breeze
sweet	woven

Words to Know

basket	small
care	mysterious
special	serene
reflected	water
fountain	spring
simple	bench
blooming	growing
lovely	arch
bright	proud

Cici loves growing colorful flowers. She grows them in a garden.

At first, the garden is empty of flowers.

O

ZERO

zero

zero

Zero.

There are zero flowers.

Cici uses a spade to dig holes.

She uses her fingers to plant the seeds.

Then Cici waits and waits

...and waits.

Finally! The plants begin to grow.

Then Cici counts her flowers.

1

ONE

one

one

One sunflower.

One **yellow** sunflower.

One tall **yellow** sunflower.

One tall **yellow** sunflower with tasty seeds.

2

TWO

two

two

Two hibiscus.

Two **blue** hibiscus.

Two giant **blue** hibiscus.

Two giant **blue** hibiscus grow as large as dinner plates.

3

THREE

three

three

Three petunias.

Three **violet** petunias.

Three sweet **violet** petunias.

Three sweet **violet** petunias dance in the breeze.

4 FOUR

four

four

Four pansies.

Four **green** pansies.

Four small **green** pansies.

Four small **green** pansies in a woven brown basket.

5

FIVE

five

five

Five orchids.

Five **black** orchids.

Five mysterious **black** orchids.

Five mysterious **black** orchids need special love and care.

6

six

six

Six lotus.

Six **pink** lotus.

Six serene **pink** lotus.

Six serene **pink** lotus reflected in the water.

7

SEVEN

seven

seven

Seven tiger lillies.

Seven **orange** tiger lillies.

Seven bright, **orange** tiger lillies.

Seven bright, **orange** tiger lillies growing near the fountain.

8

EIGHT

eight

eight

Eight irises.

Eight **purple** irises.

Eight proud **purple** irises.

Eight proud **purple** irises blooming in the spring.

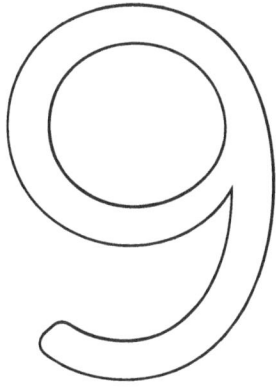

9

NINE

nine

nine

Nine daffodils.

Nine white daffodils.

Nine simple white daffodils.

Nine simple white daffodils by the garden bench.

10

TEN

ten

ten

Ten roses.

Ten **red** roses.

Ten lovely **red** roses.

Ten lovely **red** roses growing on the garden arch.

Cici loves counting her flowers.

Which flower is your favorite?

Treasure Hunt

Can you find these items in Cici's garden?

frog

butterfly

earth worm

spade

redbird

mouse

inchworm

dragonfly

goldfish

ladybug

seed packet and watering can

Cici's Super Fantastic Math Secret

If you can count forward by ones, you can add by ones.

$$0 + 1 = 1$$

one

$$\begin{array}{r} 0 \\ + 1 \\ \hline 1 \end{array}$$

one

two

$$\begin{array}{r} 1 \\ + 1 \\ \hline 2 \end{array}$$

$$1 + 1 = 2$$

one two

three

$$\begin{array}{r} 2 \\ + 1 \\ \hline 3 \end{array}$$

$$2 + 1 = 3$$

one two three

four

$$\begin{array}{r} 3 \\ + 1 \\ \hline 4 \end{array}$$

$$3 + 1 = 4$$

4
one two three four
+ 1
five
_____
5

4 + 1 = 5

5
one two three four five
+ 1
six
_____
6

5 + 1 = 6

6
one two three four five six
+ 1
seven
_____
7

6 + 1 = 7

7
one two three four five six seven
+ 1
eight
_____
8

7 + 1 = 8

8

one two three four five six seven eight

+ 1

nine

—————

9

$$8 + 1 = 9$$

9

one two three four five six seven eight nine

+ 1

ten

—————

10

$$9 + 1 = 10$$

Naomi Glasarth draws upon her vivid imagination to craft books and stories for children. From colorful counting books for the very young, to engaging high-low science fantasy, she sets her imagination free to run wild and invites you along for the crazy ride.

When not crafting her next exciting adventure, Naomi can be found enjoying the sunshine of central Arizona, working with adults with special needs, or trying to convince her body that Child's Pose is as comfortable as everyone else makes it look.

Danh Tran is a freelance illustrator with more than five years' experience and has worked on a wide variety of projects including children's book illustration, video and board games, and cartoon animation.

His happiness is in bringing ideas to life.

For more examples of his work, you may visit him at http://danhtranart.blogspot.com.

He can be reached via email at danhtran.artist@gmail.com and Skype at danhtran.artist

Hello bright and beautiful butterflies. Thank you for reading *Cici Counts by Ones*.

If you enjoyed counting forward from one to ten with Cici and her garden full of flowers, you may be interested in our other books.

Please visit us at www.burningbutterflybooks.com where we ignite imaginations one story at a time.